Toxic Algae:

How to Treat and Prevent Harmful Algal Blooms in Ponds, Lakes, Rivers, and Reservoirs

by Christopher Kinkaid

ardyne.com

Solardyne.com

Published by Solardyne, LLC
Portland, Oregon

ISBN-13: 978-1505640052
ISBN-10: 1505640059

Table of Contents

Preface

Toxic Algae is an epidemic threatening natural waterways locally, and globally. Harmful Algal Blooms (HAB) are an explosive growth of photosynthetic organisms in a body of water which can harm amphibians, fish, wildlife, pets, and human health by producing dangerous toxins.

Fresh water systems such as ponds, lakes, rivers, and reservoirs are under assault from many directions increasing chemical imbalance stressing the aquatic water systems ability to stay in healthy balance.

Toxic Algae in Harmful Algal Blooms (HAB) are an increasing threat to waterways throughout the country, and are reported with greater frequency in all 50 states. The water quality of our Ponds, Lakes, Rivers and Reservoirs are of prime importance to local communities, and the general health of local environments and ecosystems globally. Used for recreation, and as a food source for millions of people worldwide, healthy waterways are vital for ecosystems and people everywhere.

Cyanobacteria are a photosynthetic bacteria which is normally a very productive part of the aquatic ecosystem. However, when Cyanobacteria "break bad" then very harmful toxins can be produced by some species which can be highly dangerous.

Stressed from nutrient loading events such as a storm run-off, natural water bodies are being stressed to the point of imbalance introducing invasive species - including Toxic Algae.

Cyanobacteria (called a Blue-Green Algae) are a type of photosynthetic bacteria. When a pond, lake, stream, river, or other water body experiences a "nutrient loading event" conditions can trigger an algal bloom which may result in a Harmful Algal Bloom (HAB).

When nutrient levels increase, especially Phosphorous, changes in Temperature, pH, and solar energy inputs can trigger an Exponential growth in a waterbody's dominate algal species. After a rain event which washes Phosphorous into a pond, lake, or other waterbody the dominate algae will likely bloom being best positioned to adsorb the inorganic nutrients suddenly available. As the "nutrient event" was short lived the Algae which bloomed will die off as the relative nutrients have been absorbed. This die-off of the first Algal Bloom increases the amount of organic material in the water triggering a "cascade" of effects.

Aerobic aquatic bacteria which live in the water column now feast on the influx of organic material causing a large draw of Oxygen out of the water.

Now the stage is set for a Harmful Algal Bloom (HAB) to set in possibly growing an unwanted Cyanobacteria which can produce serious toxins.

There are several defenses which can be implemented to prevent the water conditions from forming which allow Harmful Algal Blooms (HAB) to occur. Climate change is evident in all 50 states measured by the increase in frequency, and toxicity of Harmful Algal Blooms spanning ponds, lakes, streams, rivers, wetlands, and reservoirs.

Higher average temperatures, increased storm activity washing nutrients into waterbodies, extended periods of drought, lower Snowpacks in the watershed all stress natural waterways. Add man-made increases in industrial and agricultural runoff and land use combined with invasive species migration and all of these factors conspire to compromise your waterbody, and the fragile aquatic ecosystems they support.

This book examines the causes of Harmful Algal Blooms (HAB) in fresh water systems, and explores methods used to combat the triggers of HAB.

Toxic Algae represents a real threat to waterways everywhere. Active assessment, planning, and action are the key to protecting your waterbody.

As climate change, population growth, and intensified agriculture and animal feeding operations grow the potential for increased frequency and toxicity of Harmful Algal Blooms (HAB) increases exponentially, and represents a real danger to water bodies, aquatic ecosystems, wildlife, livestock, pets and people everywhere.

This book examines how to treat and prevent Harmful Algal Blooms (HAB) providing a means of protecting your pond, lake, river, or reservoir from Toxic Algae.

About the Book

This book is written as a resource for treating and preventing Harmful Algal Blooms (HABs) in Ponds, Lakes, Streams, Rivers, and Reservoirs.

Toxic Algae is a condition where unwanted species of algae or Cyanobacteria invade a natural water way and produce potent toxins which are dangerous to amphibians, fish, wildlife, livestock, pets and people.

When conditions become unbalanced in waterways Harmful Algal Blooms (HAB) can form rapidly and cause catastrophic conditions for a given food web. Harmful Algal Blooms (HAB) can produce highly potent toxins which are dangerous to wildlife, farm animals, pets and people causing in some cases severe reactions, liver and brain damage, or death.

Toxic Algae is defined by the types of toxins which are produced, and their effects on organisms including humans. There are over 40 known species of Cyanobacteria (Blue-Green Algae) which can produce toxins out of thousands of individual species. These top 40 species represent a serious threat to life within your pond, lake, stream, rivers, and reservoirs when conditions occur which promote these dangerous strains.

Chapter One: Algae the Big Picture examines the vital role of Algae in the food web as a primary producer of basic organic molecules vital as nutrients for higher life forms.

Chapter Two: Water Quality discusses the various chemical conditions considered when monitoring and assessing the water conditions of a pond, lake, river, or reservoir. Defining water quality conditions is step one in beginning to quantify the health of the waterbody. Chlorophyll-a counts,Total Dissolved Phosphorous, Total Nitrogen, Turbidity, pH, dissolved Oxygen, dissolved CO2, Chemical Oxygen Demand, (COD), and Biochemical Oxygen Demand (BOD) are vital in assessing a waterbody's condition of state, and how those conditions change over time.

Chapter Three: Harmful Algal Blooms (HAB) are blooms of photosynthetic bacteria which produce toxins dangerous to amphibians, fish, animals, and humans. Toxins produced by certain Cyanobacteria include hepatotoxin and neurotoxins which are devastating to liver and brain function, respectively.

Cyanobacteria have evolved effective methods of surviving, and exploiting waterbodies which have been chemically compromised. Chapter Three explores the evolutionary tools used by Cyanobacteria to dominate a polluted waterway. Cyanobacteria have evolved mechanisms which give them an upper hand in competing with Algae, Diatoms, and Macrophytes.

Chapter Four: Nutrient Loading examines the sources of nutrients in your waterbody. External and Internal sources of nutrient loading are discussed focusing on Phosphorous as a driving factor.

Climate change has demonstrable effects on fresh water (and saltwater) ecosystems and drives a future of increasing Harmful Algal Blooms (HAB). Increasing average water temperatures, an increase in frequency and intensity of weather events which increase nutrient loading, smaller snowpacks combine to over load your waterways ability to cope and process the influx of nutrients. Storm frequency and intensity increase water mixing and availability of nutrients in the water column, with runoff into ponds, lakes, rivers and reservoirs setting the stage for Harmful Algal Bloom triggers.

Chapter Five: Oxygen is Key examines the role of dissolved Oxygen in an aquatic ecosystem and how dissolved Oxygen levels at different depths in a pond, lake, river, or reservoir are a prime driver of which organisms thrive and which do not.

Harmful Algal Blooms (HAB) can be triggered by dissolved Oxygen levels being lower than healthy in concert with Phosphorous overloading the natural ecosystem giving Cyanobacteria a competitive advantage over healthy "green" algae, diatoms and healthy Cyanobacteria strains.

Chapter Six: Aeration Technology describes techniques to mix water layers and provide a means of dissolving Oxygen into a waterbody. Comparing various techniques of aeration a waterbody can be treated with aeration technology to Oxygenate the bottom layers of a deeper waterbody.

Chapter Seven: Managing your Waterways discusses the general approach to managing ponds, lakes, rivers, and reservoirs. Data collection using a systematic method of data acquisition over time to measure daily, and monthly (seasonal) changes is vital. Technical analysis and data reduction of this data will inform water managers of the specific conditions which exist, and how they change in the waterbody. This chapter examines how Harmful Algal Blooms (HAB) can be prevented, and treated by first addressing nutrient sources and loading, and how aeration technology can begin to Oxygenate and mix the water column.

Toxic Algae is written to address treatment and prevention techniques to prevent and repair natural water ways which have been subject to Harmful Algal Blooms (HAB).

Climate change being both warmer and drier, with increased industrial and agricultural effluent runoff is combining with invasive species to produce highly stressed waterways. This book is written as a resource to better understand, prevent, and treat Harmful Algal Blooms (HAB) in ponds, lakes, rivers, and reservoirs.

Introduction

Harmful Algal Blooms (HAB) present an increasing danger to ponds, lakes, rivers, and reservoirs. When water bodies become "out of balance" chemically an opportunity for invasive algae and Cyanobacteria is created. Toxic Algae describes an invasion of "undesirable" algal, or Cyanobacteria species, which can produce very potent toxins if allowed to thrive in an aquatic ecosystem.

All water environments have their respective populations of species in a given aquatic ecosystem and when in balance are generally are healthy and productive. Harmful Algal Blooms (HAB) take advantage of out-of-balance water environments are can produce toxins which are dangerous to wildlife, farm animals, pets and humans.

Toxic Algae is a general term for Algal or Cyanobacteria species which can produce potent toxins. Toxicity is categorized as relatively mild, through moderate to high risk. Some species of Cyanobacteria can produce highly toxic Hepatotoxins and Neurotoxins which in low doses can cause function loss in the liver and brain, respectively. These Cyanotoxins represent a dangerous threat to waterways and local water bodies.

Pollution run-off, warming temperatures, and invasive species produce constant threats and impacts to ponds, lakes, streams, rivers and reservoirs. The aquatic food web is complex in even smaller water bodies and water managers seek to develop management plans which document, measure, and plot water quality conditions which develop into understanding and comprehensive action plans. Harmful Algal Blooms (HAB) occur when balanced conditions in your water body become unbalanced causing a prolific growth opportunity for specific Cyanobacteria which can produce potent toxins potentially causing sever impacts for wildlife, livestock, pets and people.

Congress, in 1998 began to address Harmful Algal Blooms (HAB) and Hypoxia with the Harmful Algal Bloom and Hypoxia Research and Control Act (HABHRCA Public Law 105-83) and established a task force to assess HABs. In 2004, HABHRCA was reauthorized and expanded tasking NOAA with assessing coastal and Great Lake HAB impacts, and has expanded to include all US freshwaters under the EPA.

Waterways are regulated by local, state, tribal, and federal agencies depending on jurisdiction. The EPA and Forestry Service manage most public freshwater bodies in the US. Preventing and treating Harmful Algal Blooms has become a greater priority as regulatory bodies seek to understand HAB, and how HABs can be abated.

Chapter One: Algae the Big Picture

Algae is a "force of nature" and plays a dominant role in a ponds, lakes and reservoirs, as well as slower moving parts of streams and rivers. Algal biomass is rich in Amino-Acids, Proteins, Carbohydrates, Lipids, Enzymes, Organic dyes, Anti-oxidants, Vitamins and other products which are highly valued.

Under normal conditions Algae is working as the best friend of a pond, lake, river, or reservoir taking up inorganic nutrients and producing a healthy food source for zooplankton, amphibians, and fish. As a "force of nature" algae in all bodies of water intended to support life are vital and very productive as a primary producer.

Algae may be the most important life form on Earth both in historic, and contemporary terms relative to the total amount of biomass and oxygen produced, and being the evolutionary root of all higher life. All complex life forms (including humans) on Earth evolved from single cell microorganisms. Amazing in reality, Algae today continues to be fundamental in supporting life on Earth daily as the base of aquatic food web and is vital for keeping your water body healthy, productive and populated with broad biodiversity. Algae (and all other plants including Cyanobacteria) through photosynthesis provide the two most important products from solar energy which make life on Earth possible: base nutrition and Oxygen.

Algae in general represent the "Good" the "Bad" and the "Ugly" of impacts for all living ponds, lakes, streams, rivers, and reservoirs depending on which species of Algae, or Cyanobacteria become dominant.

Green algae (Chlorophyta), Brown Algae (Phaeophyceae), Golden Algae (Chrysophyceae) and Diatoms (Bacillariophyceae), Yellow-Green Algae (Cryptophyta) and Blue-Green Algae (Cyanobacteria) are examples of the "Good" in algae as they are "Primary Producers" in the water system producing dissolved Oxygen (DO) into the water column, and building nutrients in their biomass through photosynthesis. The nutrients produced by terrestrial and aquatic plants are the base of the food web, and through Oxygenic Photosynthesis, produce the food upon which all higher life depend. In nature, algae and Cyanobacteria have incredible growth rates under the right conditions and roughly one half of the daily growth of algae and Cyanobacteria is consumed by "primary consumers" which support the entire aquatic food web.

Each year photosynthesis "fixes" approximately 105 Peta-grams (10 to the15th grams) of Carbon about half from the land, and half from the oceans each year. This is roughly 100 thousand million metric tons of Carbon "fixed" into sugar daily.

Cyanobacteria are our greatest friends historically, however, when Cyanobacteria go "breaking bad" then serious toxins can be produced with dangerous consequences.

The "bad" and the "ugly" of the Cyanobacterias are the relatively few species which produce potent toxins which are extremely dangerous to aquatic and terrestrial ecosystems.

Harmful Algal Blooms (HAB) are of focus because over 40 species of Cyanobacteria are known to produce dangerous toxins which can cause rash, vomiting, diarrhea, and death in wildlife, pets and humans in severe cases.

Note: Technically, Cyanobacteria are not true Algae. Bacteria are "Prokaryotic" life which involves cells which have no nucleus, or organelles for example. Algae are "Eukaryotic" which have a nucleus where DNA is stored, with defined organelles within the cell. DNA is circular with Prokaryotic cells, and Linear in Eukaryotic cells. However, Cyanobacteria are popularly known as "Blue-Green Algae" and as such will be done so here.

Over the last 2,500 million years photosynthesis evolved independently on ancient Earth at least five different times and pathways resulting in a wide diversity of Algae, Diatoms, and Cyanobacteria and the mechanisms they employ to convert selected bands of solar radiation into incredibly diverse biomass. There are eleven "divisions" of photosynthetic organisms (Algae, Diatoms, and Cyanobacteria).

Cyanobacterias are one of the most successful and important photosynthetic organisms with one of the longest genetic trails of all life on Earth.

Ancient Cyanobacteria emitted Oxygen in vast amounts using the molecule Chlorophyll-a to capture solar energy bands and drive photosynthesis. These early photosynthetic organisms transformed the early Earth's atmosphere from a chemically "reducing" atmosphere (an abundance of free-electrons) to an "oxidizing" atmosphere (an absence of free electrons) making life as we know it possible.

Converting water, sunlight, mineral salts and Carbon Dioxide (CO_2) using Chlorophyll-a fixing Carbon from CO_2 into glucose produced fast growing biomass releasing vast amounts of Oxygen into the early Earth's atmosphere.

The Great Oxygenation Event often called ultimately enabled all higher life forms in aquatic and terrestrial plants, zooplankton, nematodes, crustaceans, insects, amphibians, fish, birds, animals and ultimately humans to evolve, develop and thrive. We owe life on Earth to Cyanobacteria.

As Cyanobacteria have been a force of nature from the beginning of life on Earth it remains a force of nature today.

Ponds, lakes, streams, rivers, and reservoirs worldwide face enormous stresses and threatens waterbody health throughout the country.

Chapter Two: Water Quality

Water quality is the heart of any aquatic system and involves the balance of many factors and influences. In this chapter we'll look at the major water quality factors and indicators, and how these conditions can be influenced to bring a pond, lake, stream, river, or reservoir back into balance.

A theme which will be examined in this book is the dominant role of dissolved Oxygen in the water, and how Oxygen levels determine which life forms will be supported by your water body, and which will not.

Water flow and circulation is a critical factor in water ways. Rivers are usually moving water and provide some aeration and mixing, with eddies of calmer water in side channels. Ponds, Lakes, and slow streams have much less water movement with some Ponds nearly static. Wind upon a Pond, or Lake creates some oxygenation and mixing especially effective in shallow waters, however, these are usually not enough to keep your water bodies oxygenated at levels which support higher life such as algae, aquatic plants, amphibians, and fish when a "nutrient load event" occurs such as with a rain storm runoff.

When Ponds, or Lakes are not mixed properly then a stratification occurs isolating different layers of water within your water body at different temperatures, and different amounts of dissolved Oxygen (DO) and nutrient loading. Lack of sufficient mixing between the water layers at different depths causes an "anaerobic" condition to begin at the lower levels where dissolved oxygen critical for higher life forms becomes nearly non-existent. When deeper levels of a Pond, or Lake become anaerobic, (also called anoxic) with a severe lack of dissolved oxygen then a multiplicity of runaway effects begin to emerge.

Bottom sediments on Lake bottoms when in an Anoxic condition begin to release metals, and gasses into the water column which further exacerbate the conditions required for healthy life systems.

These bottom sediments can release manganese and iron ions, as well as hydrogen sulfide gasses which are toxic to higher life forms, as well as causing unpleasant odors.

Further exacerbating the chemical stresses on Ponds, Lakes, and Rivers, the release of Phosphorous from sediment layers in the lake bottom collect and build. The Benthic bottom layer of a pond and lake is also called the "hypolimnion" and is a vital area for your water bodies health. When Phosphorous is released under these anoxic conditions partial mixing causes Phosphorous to move into other layers of the water column causing imbalances of Nitrogen to Phosphorous rations and becomes a direct threat to zooplankton, amphibians, and fish.

Changing ratios of chemicals, as well as the chemicals themselves are some of the most sensitive and reactive conditions your lake, river, or pond will face. As we'll see later the industrial and agricultural run-off pouring into local ponds, lakes, and rivers causes massive Phosphorous and Nitrogen "loading" events. Loading is used to represent not only the Chemical Oxygen Demand (COD) this causes, but also includes the Biological Oxygen Demand (BOD) your pond experiences when subjected to pollution run-off.

The majority of Phosphorous loading for a pond or lake comes from external sources such as water fowl, the atmosphere, septic systems, household products, and industrial/agricultural chemical loading from watershed run-offs.

Ponds, Lakes, Streams and Rivers are under constant assault from these Nitrogen and Phosphorous sources which lead to dramatic impacts on water quality, and the health of your water way's aquatic species. The precursor of Harmful Algal Bloom (HAB) events.

Water Quality Factors:

Water quality is vital to the health of your pond, lake, stream, or river. The major factors which determine which life form can exist in your water body are dissolved oxygen (DO), dissolved Carbon-Dioxide (DCO2), pH, and Chemical Loading fueling Chemical Oxygen Demand (COD), and Biological Oxygen Demand (BOD).

Other physical thermodynamic factors which greatly effect how life functions in water ways is the temperature, and mixing rates which greatly influence the COD, and BOD. The turbidity of water relates to how many particulates are dissolved and have additional influences by interrupting sunlight from reaching deeper layers of water.

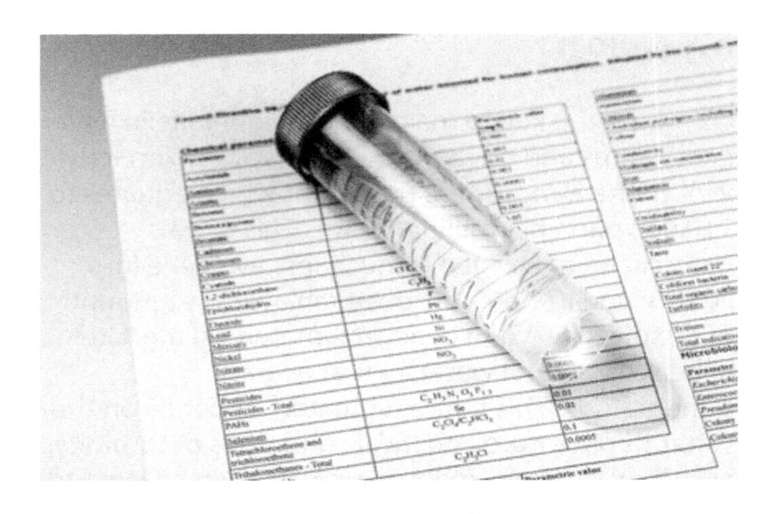

The dominate factors affecting your water quality can be quantified and analyzed. Water Quality is of great concern for water managers who wish to understand the "conditions of state" which exist for a given water body, at a given time, and through the seasons.

The most important factors in water quality include Dissolved Oxygen (DO), Dissolved CO2, pH, temperature, turbidity, Total Dissolved Solids, Chemical Oxygen Demand (COD), Biochemical Oxygen Demand (BOD),and the Nutrients (primarily Phosphorous) available for plant uptake.

The Total Maximum Daily Load (TMDL) is an important metric used to define the upper limit a water body can be loaded. Water managers often use TMDL to conceive and implement water management plans.

Oxygen (O2)

Oxygen is the key to maintaining higher life in water ecosystems and on land. Oxygen is approximately 21% of the atmosphere by weight which equates to 210,000 parts per million (ppm). This large percentage of Oxygen is much less available in liquids, and in water produces an average amount of dissolved Oxygen of approximately 10 mg/Liter in healthy natural waterways at average temperature. This figure ultimately depends on the water temperature and local conditions of turbidity, and "loading" inducing a Chemical Oxygen Demand (COD), and Biological Oxygen Demand (BOD). Note: also called Biochemical Oxygen Demand (BOD) both contribute to a Total MaximumDaily Load (TMDL).

The Total Maximum Daily Load (TMDL) will be the central metric which describes the Maximum Phosphorous and total loads which help define the water body's condition of state. TMDL levels are used by regulatory bodies such as state, and federal agencies as a measurable level which can be used to determine whether a waterbody is in compliance with regulatory affairs and regulated standards.

Oxygen usually dissolves in water at standard pressure at 14.6 mg/Liter at 0 degrees Celsius.

At a higher temperature such as 20 degrees Celsius water is unable to hold dissolved Oxygen as easily, and in solution, able to hold only 7.6 mg/Liter.

Most fish require at least 6 mg/Liter of dissolved Oxygen so the ability of water to hold DO makes life in aquatic ecosystems possible. However, if temperatures in the waterbody begin to exceed 30 degrees Celsius, then Oxygen has much more difficulty staying soluble in water and begins to threaten the minimum Oxygen required to support fish.

Dissolved Oxygen (DO) levels are critical for aquatic life at all trophic levels, and for different depths of your water body. Dissolved Oxygen (DO) levels of at least 6 mg/Liter are required to support most fish species. When DO levels drop to 3 mg/Liter then crustaceans, and amphibians are threatened. When DO levels fall below 3 mg/Liter then Dead Zones begin to form for Aerobic organisms. The water at deeper levels in your waterbody become anoxic and only able to support Anaerobic bacteria.

Anaerobic bacteria (without Oxygen) begin to thrive feeding on organic material at the bottom of your waterbody expelling further wastes into the water column including Hydrogen Sulfide which is toxic to higher aqueous life forms.

Benthic organisms which inhabit the bottom layers or burrow into the sediments including Worms and Clams require a minimum of 1 mg/Liter of dissolved Oxygen. If DO levels drop below 1 mg/Liter then Anaerobic conditions exist and the water layer becomes anoxic.

Water's lack of ability to hold onto dissolved Oxygen as water temperature rises follows Henry's Law which dictates how soluble gasses interact with liquids. Warmer water can't hold onto dissolved Oxygen (DO) as readily as dissolved Carbon Dioxide (CO_2).

Dissolved CO_2 has greater mass than dissolved Oxygen and less likely to evaporate. Water's inability to hold dissolved Oxygen in warmer water increases the natural hypoxia of the water and can occur in shallow water bodies like ponds and lakes especially in the bottom hypolimnion layer.

Carbon-Dioxide (CO_2)

Dissolved Carbon-Dioxide (DCO_2) in the waterbody is another critical factor which dictates which biological processes are stimulated and others abated.

Dissolved CO_2 is vital for aquatic based photosynthesis as one of the "feedstocks" required in addition to mineral salts, water, and photon energy (solar energy) to drive photosynthesis.

Levels of Dissolved CO_2 play an important role with pH in a water environment. Carbon Dioxide dissolved in water interacts with water with some molecules reacting to form a weak Carbonic acid (H_2CO_3). This weak mineral acid helps to lower pH, or keep pH within the 6.5-7.8 level considered normal for ponds and lakes.

Algal and Harmful Algal Blooms (HAB) absorb dissolved CO_2 in the water for photosynthesis resulting in a pH rising (also due to metabolites produced by algae).

Cyanobacteria are very successful with their ability to tolerate higher and higher pH alkalinity in ponds, lakes, and slow moving areas of streams and rivers. This higher tolerance provides a competitive advantage over Diatoms and Green algal species.

Eutrophic conditions from a "Nutrient Loading Event" whether external or internal in source produces a spike in available Phosphorous and therefore algal growth rates. Harmful Algal Blooms (HAB) with rapid photosynthetic activity exacerbate the pH issue of rising alkalinity by combining to pull CO_2 out of the water column.

Dissolved CO_2 levels can be estimated with a Hatch Field Test Kit using a NaOH titration method with a phenolphthalein indicator.

Turbidity

Turbidity is the level of dissolved solids and suspended solids in a water sample. A high turbidity is a cloudy water which is mostly opaque. High turbidity suggests nutrient overloading could also be occurring.

Measured often as Total Suspended Solids (TSS), or Total Solids (TS) turbidity affects Algal and

Cyanobacteria growth by interfering with solar energy traveling through the water column. Turbidity occurs from mechanical agitation from rain, wind, biological sources, bubbles from sediments, and from disturbances from animal and human activity causing a mixing of water layers and particles both organic, and inorganic.

Turbidity in a pond, lake, or river suggests that Internal sources of Phosphorous nutrient loading are occurring which also spur algal and Cyanobacteria growth. Turbidity can be measured with a turbidimeter such as a Hach model 2100P.

pH

The pH of a water environment is dominant in dictating which life forms can thrive, and which cannot. Most freshwater systems "prefer" a pH of slightly alkaline from 7.2 to 7.8 pH. In this pH range Algae, Cyanobacteria, Zooplankton, Amphibians, and Fish seem to thrive in greatest abundance.

During algal bloom events and Cyanobacteria Harmful Algal Blooms (HAB) pH tends to be higher and can quickly rise to over 9 in shallow or smaller water bodies. Aeration, agitation, and fountains are a method employed to Oxygenate the water. Aeration or agitation also dissolves CO_2 from the air which partially reacts with water producing a weak Carbonic acid which helps to lower pH. However, during algal blooms and subsequent die-off dissolved CO_2 in the water increases dramatically as anaerobic bacteria also expel CO2 during decomposition.

Chlorophyll-a

Chlorophyll-a is a protein which is arguably the most valuable molecule on Earth. Photosynthetic organisms using Chlorophyll-a are remarkable for absorbing very specific wavelengths in the solar spectrum to initial an energy "cascade" resulting in an electron transport chain which begins the process of photosynthesis. Chlorophyll-a absorbs peak wavelengths in the Orange-Red (665 nm), and Violet/Blue (445 nm) segments of the solar spectrum to set in process the energy to "oxidize" water "harvesting" electrons and a proton which it keeps separated, producing an electron gradient. Oxygen is discarded at his point as a waste product.

Nearly all photosynthetic organisms use Chlorophyll-a as their primary pigment. And, many different kinds of phototrophic organisms Algae, Diatoms, Cyanobacteria, and Macrophytes (large aquatic plants) also have evolved other proteins which absorb specific wavelengths present in the solar spectrum. These Secondary, or "accessory" pigments allow the plant (both aquatic and terrestrial) to capture and utilize more of the available wavelengths in Sunlight unused by the Primary Pigment (Chlorophyll-a).

Solar energy represents a wide range of electromagnetic-radiation.

In photosynthetic organisms the range of solar energy available, in whole, to drive photosynthesis spans the ultra-violet at 400 nm in wavelength to the Red end of longer wavelengths at 700 nm.

Solar energy has a very broad range of wavelengths when radiated by the sun, however the Earth's atmosphere absorbs the very short, and the very long wavelengths leaving a "window" of Photosynthetic Active Radiation (PAR) between 400-700 nm in wavelength.

Chlorophyll-a has peak absorption in two very distinct and narrow wavelengths present in sunlight. The first absorption peak occurs in the violet-blue (465 nm) region, with the second peak in the Orange-Red (665 nm).

"Green" plants appear Green to the human eye because Chlorophyll-a doesn't need, or readily absorb Green wavelengths of light. Green light being largely reflected, or transmitted from the plant to the human eye makes Chlorophyll-a appear Green.

However, the real photochemical action of Chlorophyll-a is in the specific wavelengths (reds and blues) which allow the proteins to transfer "resonance" to very specific Chlorophyll-a molecules which are found in the "photo reaction centers" (P680 P700) which begin the "electron-transport" chain driving the process of photosynthesis.

Interestingly, these absorption "peaks" are very narrow and only represent a small fraction of the energy available in sunlight which makes it to the surface of the Earth.

Cyanobacteria have evolved a select group of additional light absorbing (phototrophic) proteins called Phycobilliprotiens with a very special sub-set of proteins (phycocyanin) which increase the photon energy captured at low light levels. Cyanobacteria have evolved mechanisms to deal with a wide range of "stresses" other autotrophs would find difficult to survive including these accessory pigments to augment specific photon capture, energy transfer, and utilization.

Measuring Chlorophyll-a gives a direct relationship to the amount of living biomass within a waterbody.

Chemical Oxygen Demand (COD)

Chemical Oxygen Demand (COD) measures everything in the water column (organic or inorganic) which can be chemically oxidized. Metal ions in the water react with the Oxygen and become oxidized for example, which pulls, or "demands" dissolved Oxygen levels out of the water lowering the available Oxygen for the food web. Orthophosphate (reactive Phosphorous) in the water column can be determined using Hach method 8084 (Hach, 2000). Measurements using USEPA method 410.4 is often referenced for standard measurement protocols.

Biological Oxygen Demand (BOD)

Biological Oxygen Demand (BOD) is more specific than COD as BOD is concerned with respiration of organisms including aerobic bacteria, algae, diatoms, Cyanobacteria, Macrophytes, zooplankton, worms, amphibians, and fish.

Biological (or Biochemical) Oxygen Demand (BOD) signals a draw of Oxygen out of the water column being consumed in the living organisms of the waterbody including all aquatic aerobic bacteria, plants, amphibians, and fish. During Harmful Algal Blooms (HAB) induced die offs of competing aquatic plants, and fish aerobic bacteria can draw enormous amounts of Oxygen out of the water. During the night, when plants are respirating, Oxygen is drawn out of the water to power metabolism, reproduction, and biomass accumulation. BOD measurements can be done using Standard Method 5210B.

When ponds, lakes, rivers, and reservoirs become chemically imbalanced the remarkable mechanisms of Cyanobacteria can take hold, and the potential exists to trigger a Harmful Algal Bloom (HAB) potentially producing Hepatotoxins and Neurotoxins which pose an extreme danger to public health.

Aeration is an effective non-toxic, and non-invasive method of repairing the damage caused by nutrient loading, high alkalinity, and hypoxia in your waterbody. The deeper the aeration units are placed in your waterbody the more water is moved with the effect of "turning over" the entire water volume over time.

Turning over your water brings hypoxic water from the bottom to the top, with Oxygenated surface water sinking to replace the water lifted. This "turn over" rate will be the most effective way to transfer Oxygen into your waterbody, supporting the most important aspect of your water quality health: dissolved Oxygen.

Measurements of color, Nitrogen and Phosphorous levels can be accomplished using a Hach DR/2010 spectrophotometer. Other protocols include the USEPA Method 365.2 and Standard Methods for the Examination of Water and Wastewater (Eaton, 2005).

Mapping your waterbody depth and location can be captured using sonar mapping gear such as a Lowrance HDS5 fish finder/GPS chart plotter to record depth and location data.

Shoreline survey data can be collected using a Magellan Triton 400 handheld GPS. Software such as SonarViewer can be used to convert Lowrance data into images.

Chapter Three: Toxic Algae
Harmful Algal Blooms (HAB)

Toxic Algae are typically those species of Cyanobacteria which produce Dermal Irritants, Hepatotoxins, and Neurotoxins which are extremely dangerous if allowed to bloom unabated.

However, the term "toxic" must be applied with a matter of perspective. There is an old axiom in Medicine: "Dose makes the poison." The relative "amount" of toxins really define the impacts.

The extreme danger of Harmful Algal Blooms (HAB) is the incredible potency of these toxins capable of toxic impacts with relatively small doses presenting a grave danger to wildlife, livestock, pets and people.

Lake managers have employed a series of steps to first assess a Harmful Algal Bloom by assessing the water quality conditions, then implement a plan to treat, and/or prevent Harmful Algal Bloom (HAB) events.

Lake managers seek to employ "surveillance" of the pond, lake, stream, river, or reservoir to measure conditions in the water, and as dynamic systems which are multivariable and ever changing, a measurement system over time should be established. Once a "condition of state" and how those conditions change through the seasons, can be established then prevention or treatment plans can be formulated and implemented.

Algae can be the "best friend" of your water way, or your worst enemy if chemical factors go out of balance and undesired Cyanobacteria can take hold.

There are millions of algal, and Cyanobacteria species in the world with only a small fraction known to produce toxins. However, the toxins which can be produced by some Cyanobacteria are incredibly potent and dangerous for fish, wildlife, livestock, pets and humans.

Harmful Algal Bloom (HAB) events have been increasing with some taxa of Cyanobacteria able to produce highly potent toxins such as Microcystin, a

Hepatotoxin affecting the liver, and Anatoxin-a which is a Neurotoxin which is severely damaging to the brain.

Harmful Algal Blooms (HAB) have driven research on algal strains with an emphasis recently on specific photosynthetic bacteria strains of Cyanobacteria due to these potent toxins.

Algae, and Cyanobacteria grow wildly when conditions allow with species capable of adapting and thriving in a wide range of conditions.

Algae, Diatoms and Cyanobacteria "battle it out" in all the water ways in the world from mud-puddles to the oceans, competing for available nutrients and solar exposure.

Each have developed effective "methods" of edging each other out such as forming "mats" or "filaments" which serve to block light from other competing species. With millions of different species of Algae and Cyanobacteria found all over the world each species has become well adapted and specialized honing in on its "preferred" water environment.

Cyanobacteria have a long history from ancient Earth stretching back 2.5 billion years in the fossil record demonstrating an epic achievement in forging life on early Earth. Cyanobacteria are some of the most successful organisms in the vast history of life on Earth having experienced a wide range of conditions from "Ice Ball Earth" to massive volcanic events throughout millions of years, enduring extreme geophysical events throughout the Earth's history, and have the genetic diversity to show it. In ponds, streams, lakes, wetlands, rivers, and reservoirs Cyanobacteria can go "toxic" when conditions allow these incredible engines of life to "break bad."

Specific Cyanobacteria capable of producing toxins take hold and begin to "out compete" other aquatic life for available nutrients and solar energy once they are "triggered" by water becoming "eutrophic" with nutrient loading. Once the Nitrogen to Phosphorous (N:P) ratios start to drop from a preferred 18-22 down under 10-12 then conditions become favorable for an invasive Cyanobacteria.

Eutrophic conditions in your waterbody trigger a rapid growth response as Cyanobacteria have an incredible and opportunistic appetite and have evolved to grow rapidly when they can, to gain energy for when they can't.

Cyanobacteria have a secret life which includes some of the most fantastic tools for adaptation to rapidly changing, and extreme environments encountered in the natural aquatic world.

These evolved emergency response mechanisms have enabled them to thrive over deep time, developing "machinery" for example to capture Nitrogen directly from the air where most plants on Earth can't do that trick.

Most plants gain their vital Nitrogen compounds from dissolved sources in aquatic plants, or from their root system in terrestrial plants.

Cyanobacteria naturally occur and are found all over the Earth in fresh water, brackish water, salt water, soil, ice, air., and even inside porous rocks. Cyanobacteria are found throughout the lithosphere and hydrosphere at all altitudes, climatic zones, and latitudes.

As arguably the most successful organism on Earth Cyanobacteria have evolved an impressive "tool kit" for adapting and coping with conditions most Green Algae and Diatoms would find hostile. This remarkable robustness of Cyanobacteria make them a formidable force when facing a Harmful Algal Bloom (HAB) event.

Evolved from early Earth where conditions of pH, temperature, high UV light levels and other stresses would make any condition for life highly unlikely Cyanobacteria found a way to take fundamental elements of selected wavelengths in sunlight, water, CO2 with trace inorganic salts, and using Chlorophyll-a, produce the most important process on planet Earth: Oxygenic Photosynthesis.

Oxygenic Photosynthesis is the process which produces daily the base nutrition and Oxygen needed by all aquatic and terrestrial life on Earth.

Cyanobacteria is a warrior with a long track record surviving and thriving in harsh conditions which on the very early earth existed without the presence of much Oxygen. The Oxygen produced by early Cyanobacteria as the first "waste product" enabled the evolution of all higher life forms by producing enough Oxygen to support higher cell formation and function.

Cyanobacteria produced the original toxin (Oxygen), and is responsible for the Great Oxygenation Event of early earth leading to all other aquatic, then terrestrial plants, then on to evolve all zooplankton, crustaceans, amphibians fish, insects, birds, animals and ultimately humans on Earth.

Cyanobacteria is an ancient and successful order of life having evolved unusual and effective mechanisms which yield many advantages in competition for resources with other aquatic species in the real world when conditions become stressed.

Cyanobacteria's diverse abilities to control its position in the water column, as well as the incredible "biotechnology" which allows Nitrogen to be absorbed from the air by many Cyanobacteria strains (a trick most plant's on earth can't do) give Cyanobacteria a formidable edge in competing for nutrient resources and solar exposure with other aquatic species.

Cyanobacteria Advantages:

Cyanobacteria have many advantages evolved to give the organism a competitive advantage to survive a changing and often harsh environment.

Buoyancy and Position in water column

Many Cyanobacteria taxa have evolved an advantage living in the water with the formation of internal gas vesicles which allow them to control their position in the water column by adjusting their buoyancy. The evolutional advantage gives Cyanobacteria an ability to seek the brighter light available near the surface by inflating and rising in the water column. This advantage gives Cyanobacteria a means to out-compete other Algae and Diatom species which are unable to control their depth as readily.

Nitrogen Fixation

All plants require inorganic forms of Nitrogen in forms which can be absorbed including Nitrates, and Nitrites. Many Cyanobacteria (about 1/3 of known Cyanobacteria) have the ability to "fix" nitrogen directly from the atmosphere and produce Nitrates and Nitrites which are bioavailable for the Cyanobacteria to synthesize complex acids, proteins, and enzymes.

Cyanobacteria use highly specialized cells called Heterocysts which utilize a special enzyme

(Nitrogenase) to fix nitrogen into complex molecules, enabling an extreme advantage compared to most algae which can only access Nitrogen in dissolved compounds which are waterborne and may be limited.

Although, available Phosphorous is always the limiting factor for Cyanobacteria in a common aquatic environment, the ability of certain Cyanobacteria to use atmospheric Nitrogen when dissolved inorganic Nitrogen is unavailable or limited, demonstrates Cyanobacteria ability to metabolize Phosphate as Nitrates and Nitrites which are required by the organism to process the Phosphoric compounds.

Phosphorous -

Phosphorous is the dominant nutrient which dictates algal growth rates. Cyanobacteria under usual aquatic ecosystem conditions are not particularly effective at competing for Phosphorous with other Algae and Diatoms in Fresh water systems when N:P ratios are "normal" around 16-22 in magnitude. Under these "generally healthy" conditions Cyanobacteria are rarely dominant as an aquatic species which makes Harmful Algal Blooms rare or non-existent in a properly balanced N:P water environment.

When a nutrient loading event occurs in a water way, algal blooms may become evident. A Harmful Algal Bloom is a direct indicator that external nutrient loading, temperature, pH, and internal loading (sources of Phosphorous) are out of balance. When water conditions tighten up with the external or internal loading of Phosphorous into the water then the "door opens" for Cyanobacteria rapid growth and proliferation prompting a Harmful Algal Bloom event.

When Phosphorous is limited (not easily available) in a water body, Cyanobacteria have a difficult time growing and competing with other aquatic plants for nutrients and sunlight. When Phosphorous is not limited (readily available) and increases due to nutrient loading then Cyanobacteria can thrive and enter an exponential growth phase potentially resulting in a Harmful Algal Bloom.

Nitrogen:Phosphorous Ratio -

The Nitrogen to Phosphorous ratios (N:P ratio) as discussed above is a critical factor in unleashing, or controlling a Harmful Algal Bloom of Cyanobacteria. When the N:P ratio is from 18-22 then generally healthy conditions can exist in a "normal" aquatic ecosystem. During a Phosphorous loading event the relative ratio drops and when reaches the 5 or 6 level then invasive Cyanobacteria species can begin to take hold.

Optical Energy -

Cyanobacteria are some of the earliest life forms on Earth. Their photosynthetic response gives

Cyanobacteria some special evolutionary "tools" which utilize primary and accessory pigments giving them the ability to tap various wavelengths in sunlight to stimulate photosynthesis. The primary pigment of most Cyanobacteria is Chlorophyll-a the ancient cornerstone of oxygenic photosynthesis.

Chlorophyll-a is probably the most valuable molecule on Earth as nearly all aquatic and terrestrial plants use Chlorophyll-a as the most productive pigment to capture solar radiation for photosynthesis. Chlorophyll-a is most absorptive in two "bands" of solar energy sharply peaking in the Violet/Blue (445 nm) region and the Orange/Red (665 nm) parts of the spectrum. Most Cyanobacteria have evolved use of additional proteins called accessory pigments or Secondary Pigments which capture other parts of the spectrum and further drive photosynthesis in the organism.

Secondary, or Accessory pigments used by plants both aquatic and terrestrial include Chlorophyll-b, Chlorophyll-c, Chlorophyll-d, Carotenoids and specialized pigments called phycobiliproteins. Cyanobacteria use specialized accessory pigments including phycobilins a form of phycobiliprotein which responds to lower light levels giving Cyanobacteria a competitive advantage at depth, or at lower levels of solar energy from clouds, storms, and seasonal changes.

Solar energy shining on a water body follow diurnal (daily), and seasonal cycles. Cyanobacteria using Phycobilins have a competitive edge in cloudy, or inclement micro-climates using secondary pigments which increase their photon capture in low light conditions.

Anaerobic Environments -

Cyanobacteria have mechanisms evolved to cope with extreme stress. This feature makes them a difficult and entrenched organism when a Harmful Algal Bloom (HAB) event takes hold. Algal blooms resulting from rapid nutrient loading followed by Harmful Algal Blooms spur plant and fish die-offs producing a high influx of organic material suspended in the water column. This condition feeds aerobic bacteria in the water column increasing Biological Oxygen Demand (BOD).

As organic material collects in the waterbody an abundance of organic material settles and collects on the bottom where further decomposition occurs promoting anaerobic bacteria which is caustic to aerobic life and leads to hypolimnion (bottom) layer dead zones.

Resting Cells to weather the storm -

Cyanobacteria have another advantage in their response to extreme environments, or rapid changes in an existing environment. Resting cells (Akinetes) can be formed by Cyanobacteria when facing an extreme stress factor such as prolonged drought causing a pond or lake drying out. Produced from a vegetative cell the Resting Cell is larger than usual cells and produce a structure which helps protect the cell from physical damage. A sort of life raft is formed as a hollow ball exoskeleton. Historical land surveys use Soil Core samples to measure the Resting Cell (Akinetes) density to indicate past Algal Blooms and drought cycles. Resting cells give Cyanobacteria an ability to "outlast" the harsh environment and is a formidable ability to survive.

Predation -

Cyanobacteria under "normal" conditions exist in heathy water ways and are "prime producers" being important sources of dissolved Oxygen and base nutrition largely eaten by Zooplankton. There are thousands of species of known Cyanobacteria with only a relative few (40+) known which produce toxins.

Species of Cyanobacteria such as Spirulina produce a wide array of excellent amino-acids, proteins, lipids, enzymes, vitamins and anti-oxidants which are very healthy for consumption.

However, in the Case of Cyanobacteria producing Cyanotoxins, there are specific Cyanobacteria taxa which produce potent toxins, and probably evolved, as many poisonous species do, to discourage predators from eating them.

As a primary producer along with healthy Algae, Diatoms and Macrophytes, Cyanobacteria are a principal food producer for zooplankton the next step up in the trophic level of the food chain.

Cyanobacteria has also evolved a mechanism of forming large colonies, a "strength in numbers" approach to surviving being eaten by zooplankton.

Forming large matts, or surface scumming, Cyanobacteria try to discourage being eaten. Cyanobacteria are normally consumed by zooplankton which keep populations in a manageable level under normal healthy conditions.

Temperature Tolerance -

Temperatures, and changes in temperature have dramatic impacts on aquatic species which are extremely sensitive to water temperatures outside their preferred range.

Cyanobacteria have an enormous range of tolerable temperatures, including warmer temperatures over 30 degrees Celsius, which explains their rapid growth in shallow bodies of water, or surface layers of larger bodies when Summer temperatures, and Climate Change impacts bring local temperatures to higher levels.

Green Algae, and Diatoms have a narrower range of preferred temperatures and as a whole are less tolerant to temperatures over 30 degrees Celsius.

Cyanobacteria being more temperature tolerant have honed their response mechanisms to cope giving Cyanobacteria an opportunity to exploit the proper water conditions of being eutrophic, and warm.

pH determines reactions -

In all of Chemistry pH is one of the most important factors in determining whether an aqueous chemical reaction can occur, let alone a biological reaction. The pH is a measure of acidity or alkalinity of an aqueous solution. Being a logarithmic scale each number on the pH scale is either 10 times larger or smaller than another.

A pH of 7 is neutral with any number lower being more acidic. Numbers above 7 become more alkaline, a pH of "8" has 10 times the alkalinity of neutral. A pH of "9" has 100 times the alkalinity of "7."

A pH of "5" is 100 times more acidic than a pH of "7" and illustrates how even small changes in pH can have dramatic impacts on chemical and biological processes. The condition of pH is vital with Cyanobacteria showing an increased ability to tolerate High pH levels.

All of these aspects give Cyanobacteria, including potentially toxic species, a strong competitive advantage when a pond, lake, river, or other waterway becomes stressed chemically when

Phosphorous becomes more available due to external or internal sources.

The toxins produced by Cyanobacteria are classified by how they affect living tissue and specific organs in fish, animals, and humans. Cyanotoxins will most often collect in the fatty tissues and organs of an organism, but are still present in lesser levels in muscle tissue.

Cyanobacteria toxins can be Hepatotoxins affecting the LIver, Neurotoxins effecting the Brain, or Dermal irritants which affect the skins of amphibians, fish, birds, animals and humans.

Cyanotoxins present in water can expose wildlife, farm animals, pets and humans to toxicity with casual contact and particularly if inhaled, swallowed, or absorbed through mucus membranes. Even small exposures to Cyanotoxins can lead to more serious toxic effects including loss of organ function, and in severe cases death.

Note: All Cyanobacteria listed below are known to produce the Dermal Toxin Lipopoly-saccharides which can cause sever skin irritation.

Below is a List of Cyanobacteria Genera which are know to produce dangerous toxins including Hepatotoxins and Neurotoxins:

Cyanobacteria	Hepatotoxins	Neurotoxins
Anabaena	Microcystin, Cylindrospermos in	Anatoxin-a, Anatoxin-a(s), Saxitoxins, Beta-N-methylamino-L-alanine
Anabaenopsis	Microcystin	-
Aphanizomenon	Cylindrospermos in	Anatoxin-a, Saxitoxins,
Arthrospira	Microcystin,	-
Cyanobium	Microcystin,	-
Cylindorspermo psis	Cylindrospermos in	Saxitoxins,
Gloeotrichia	Microcystin,	-
Hapalosiphon	Microcystin,	-
Limnothrix	Microcystin,	-
Lyngbya	-	Saxitoxins,
Microcystis	Microcystin,	Anatoxin-a
Nodularia	Nodularins	-
Nostoc	Microcystin,	Beta-N-methylamino-L-alanine
Oscillaroria	Microcystin,	Anatoxin-a, Homoanatoxin-a

Cyanobacteria	Hepatotoxins	Neurotoxins
Phormidium	Microcystin,	Anatoxin-a,
Planktothrix	Microcystin,	Anatoxin-a, Homoanatoxin-a Saxitoxins, Beta-N-methylamino-L-alanine
Raphidiopsis	Cylindrospermos in	Anatoxin-a, Homoanatoxin-a, Saxitoxins, Beta-N-methylamino-L-alanine
Synechocystis	Microcystin,	-
Umezakia	Cylindrospermos in	-

Cyanobacteria toxins are incredibly potent often requiring only parts per billion to have impacts.

Toxic levels in water will vary based on the Cyanobacteria cell concentration with a cell count of 20,000 cells per mL being serious and the usual threshold for public health alerts. Hepatotoxins have serious effects on Liver function damaging livers in amphibians, fish, and mammals (humans) and have no effective antidotes. Neurotoxins produced by these Genera and others unclassified are also without effective antidote and present a danger to wildlife, farm animals, fish, amphibians, pets and people.

Neurotoxins, and Hepatotoxins are not effectively filtered with traditional sediment (10 micron), and Carbon (5 Micron) filtering techniques, and are unaffected by boiling the water, representing a serious public threat if Cyanobacteria is present in numbers over 20,000 cells/mL. Cyanotoxins including Neurotoxins, and Hepatotoxins can be dangerous at levels as low as 1 part per billion.

Harmful Algal Blooms (HAB) are a serious matter for many fish, animals, and humans with many water management personnel becoming more proactive in pond, lake, stream, river, and reservoir health, and management.

The World Health Organization (WHO) has been developing guidelines for water quality relative to Cyanotoxins in an effort to protect public health. When animals and humans use waterways the usual paths for exposure to Cyanotoxins include the direct contact of body parts, especially membranes such as eyes, ears, mouth, and throat.

Accidental swallowing of water and inhaling water particles all generate pathways of exposure and can cause vomiting, diarrhea, and in severe cases death. When Cyanobacteria reach cell densities of 20,000 cells per mL as measured in water samples most Public Health Divisions issue public alerts of a probability of adverse health effects.

Note: The threshold level of 20,000 cells per mL of water is not universally accepted or implemented by water managers and is only suggested. It is a recommended threshold for a quantitative estimate at acceptable levels of toxins present which should alert public health threat alerts. Cyanobacteria cell counts of 100,000 cells per mili-liter of water represent a serious threat and can reach a cell density visible as a pond scumming or matting.

These are visual indications of a serious Harmful Algal Bloom and adverse health effects are likely with contact with fish, animals and humans.

The World Health Organization (WHO) has suggested a general drinking water guideline of 1 microgram per liter for Microcystin as the maximum tolerated and still be potable (1 ppb).

Runoff from a local watershed can create external loading of Phosphorous resulting in water eutrophication. The first stage usually begins with the dominate Green algae, then rapid die-off producing a lot of organic biomass in the water. Water borne aerobic bacteria then feast on the organic material pulling Oxygen out of the water.

Chapter Four: Nutrient Loading

Phosphorous is the most important macro-nutrient which drives nutrient loading and Harmful Algal Blooms.

Ponds, lakes, streams, rivers, and reservoirs are stressed chemically from External and Internal nutrient loading sources, with Phosphorous released in each case being of most concern.

When watershed run-off enters a water body all aspects of the "aquatic conditions" are affected including pH, N:P ratios, and therefore dissolved Oxygen (DO) levels which has dramatic impacts on Oxygen demands in the water competing with fish and other aquatic plants. Dissolved Oxygen (DO) levels in water are the product of the Chemical Oxygen Demand (COD), and the Biological Oxygen

Demand (BOD) which wants to pull Oxygen out of the water.

The dissolved Oxygen (DO) levels of a given water body are paramount and must be considered for various depths of the water. The bottom layer of water (the hypolimnion layer) places a critical role for the health of your water body and is the most susceptible to losing Oxygen and becoming hypoxic. In these Anaerobic conditions Anaerobic bacteria begin to thrive with often harsh consequences for all other life in your water body. Anaerobic (without Oxygen) bacteria produce Hydrogen Sulfide and other other gasses with become toxic for amphibians, fish and animals.

Oxygen is the third most abundant element in the Universe behind Hydrogen, and Helium, respectively.

Water is 88.88% Oxygen by weight, and indeed Oxygen is the most abundant element in the Earth's crust. Oxygen in the Earth's crust has combined with metals and minerals and is found everywhere oxidized with anything which could be oxidized. Even the biosphere is dominated by Oxygen bound in chemical form as the most abundant element by weight in most living things.

Oxygen (O2) composes about 21% of the weight of the atmosphere. In water ways such as ponds, lakes, rivers, and oceans the amount of Dissolved Oxygen (DO) varies greatly and depends on the local conditions. Fish require at least 6 mg/Liter of dissolved Oxygen to be healthy.

Oxygen is very reactive. Oxygen levels in a water body reflect a Chemical Oxygen Demand (COD) which is the draw of Oxygen out of the water as Oxygen reacts with metal ions "rusting" the material and being pulled out, and a Biological Oxygen Demand (BOD) as aquatic plants and Aerobic bacteria which also live in the water column respirate and absorb Oxygen for metabolic, and reproductive function.

During the day when photosynthesis is active, levels of dissolved Oxygen (DO) increase as photosynthesis releases Oxygen into the water from the oxidation of water. During the night when there is no photosynthesis occurring, cells use the same machinery in "reverse" by absorbing Oxygen from the water to use for metabolism. Plant cells respirate at night and produce a Biochemical Oxygen Demand on the water depleting the Oxygen available. This nightly "hypoxia" cycle is tolerable for aquatic plants under normal conditions. However, when Phosphorous levels spike, prompting a bloom die-off cycle, they conspire to bring night time dissolved Oxygen levels dangerously low to the point where desired flora are killed and can't recover.

Due to photosynthesis dissolved Oxygen (DO) increases during the day, and decreases during the night in a natural diurnal cycle. If an Algal Bloom becomes a Harmful Algal Bloom (HAB) then dissolved Oxygen levels can be further decreased as algal biomass from the first event begin to decompose overloading the water column with organic material.

Aerobic bacteria, a healthy and pervasive order of life in a natural water way are "primary decomposers." As primary decomposers, aerobic bacteria in the water will also draw Oxygen from the water (as they respirate) and draw available Oxygen which help decompose organic materials. This produces a large Biochemical Oxygen Demand (BOD) on your lake or water body.

Take for example a Pond, or Lake. Let's imagine this lake is surrounded by lake home owners and a storm with a big rain event washes Phosphorous into the lake. The "good" algae, including Green, Brown, Golden Algae and Diatoms being very well adapted to adsorbing and utilizing in-organic nutrients in the runoff increases their growth rates and we have a "good" algae bloom.

Because a storm event is short lived, perhaps a few hours for example, the runoff has been abated in a short time.

The Green Algae will use the most available nutrients and blooms producing an "exponential growth rate." Algae are short lived, and starved of the recent influx of nutrients experiences a mass "die-off" event.

Now, the aerobic bacteria in your waterbody come into play and begin to feast on the algal biomass in the water causing a "good" bacteria bloom to occur which pulls dissolved Oxygen out of the water lowering dissolved Oxygen (DO) levels and causes extreme stress for the higher life forms in the water based ecosystem.

Managing these stresses to keep a waterbody healthy is the principle goal of a water manager.

To manage ponds, lakes, rivers and streams water managers seek to measure and implement a standard maximum "stress" defined by the Total Maximum Daily Load (TMDL) for a given body of water. The TMDL is a product of all Chemical and Biochemical Loads often including turbidity and total dissolved solids. TMDL usually focus on Phosphorous but can also include Nitrogen, Potassium, and other elements and compounds which present a load on Oxygen Demand in the water column.

The Management Plan will be designed to lower TMDL and implement strategies which will keep the waterbody under federally allowed TMDL levels.

Example of Limiting Phosphorous to treat a Harmful Algal Bloom:

Harmful Algal Blooms (HAB) have occurred with great frequency due to anthropogenic sources of nutrients principally Phosphorous.

Lake Washington near Seattle, WA is a well studied example of treating a large lake from suffering a Harmful Algal Bloom.

Cyanobacteria species Oscillartoria rubescens, which was plaguing the lake, was totally abated by limiting the high Phosphorous levels which were being washed into the lake from watershed run-off. By limiting the influx of Phosphorous, Lake Washington was able to stop a recurring Harmful Algal Bloom event from the Cyanobacteria Oscillatoria rubescens, and totally eliminate the threat of Harmful Algal Blooms from this taxa.

External Nutrient Loading

External nutrient loading occurs when run-off from agricultural, industrial, commercial, and residential activities pour massive amounts of Nitrogen and

Phosphorous as well as many other elements, chemicals and hormones into your pond, lake, stream, river, or reservoir. External loading not only introduces massive amounts of nutrients available for plant uptake, but also changes the "Ratios" of these elements and chemicals promoting an environment which can be exploited by Cyanobacteria.

Considering an optimum composition and ratio of nutrients for general Green plants (Chlorophyta) which include all plants which utilize Chlorophyll-a has been studied and analyzed. The Redfield Ratio refers to the general optimum ratio of important macro-nutrients and include Carbon, Nitrogen, Phosphorous, and extends to Micro-nutrients. The general Redfield Ratio of C:N:P is 106:16:1, respectively. That is 106 parts Carbon, to 16 parts Nitrogen to 1 part Phosphorous.

The Redfield ratio approximates the general ratio of macro-nutrients within biomass and sets the stage for ideal growth conditions for desirable Green Algae as a nutrient mix. As run-off brings Phosphorous into a water body the relative Ratio of Nitrogen to Phosphorous is changed and drops in numeric value. The ratio of Nitrogen to Phosphorous (N:P) is a critical factor driving plant growth rates including algae in general, and Cyanobacteria in particular, as each has their preferred range. The lower the N:P ratio of your water the more likely your pond, lake, stream, river, or reservoir will experience Harmful Algal Blooms

Green algae, and fresh water Diatoms, which normally occupy a water column is normally a healthy and important "primary producer" for introducing Oxygen into a water body, and water column. Ratios considered "normal" range from about 16-22 for Nitrogen to Phosphorous which produces the optimum conditions for Green Algae, and Diatom nutrient uptake.

However, external nutrient loading of Phosphorous into your waterway lowers this relative ratio, and introduces conditions which are more favorable to cyanobacteria taxon.

As the Nitrogen to Phosphorous ratio drops a dual effect results in "Good" Green Algae being stressed with Cyanobacteria (undesired in this case) being promoted.

External nutrient loading of ponds, lakes, streams, rivers, and reservoirs is a major driving force for Harmful Algal Blooms. Watershed run-off contains many contaminates but Phosphorous is the leading driver for chemical and biological impacts. Phosphorous is very important because usually Phosphorous is Least available macro nutrient. When Phosphorous becomes in abundance in your water body then massive impacts occur to the "preferred" balance of nutrients.

Watersheds have enormous runoffs and produce rapid imbalances with massive influxes of Nitrogen and Phosphorous dominated effluents. Agricultural, and Industrial effluents are particularly important as they can massively impact a water body in very short time. Global climate impacts have seen more frequent storms inducing increasing Nutrient Loading Events. These effluents have dramatic affects on pond, lake, stream and river eutrophication, which affects not only the water column at different depths, but also settles in sediments which can contribute to internal loading at a later date when these Phosphates can be reintroduced into the water column.

The links between Phosphorous loading and Algal Biomass are well established (Cook et al, 2005) as well as evidence which demonstrates, if you lower nutrient loading running into a waterbody you can recover pond, lake, stream, or river balance and effectively treat the condition which supports Harmful Algal Blooms.

The control of nutrient loading from external sources includes a "bottom up" approach to lake water management.

The productivity and ecosystem dynamics are a function of nutrient and energy inputs to the base of the food chain. The bottom up approach looks at the primary producers as the base of the food web and seek to manage the base as a way to manage the ecosystem. Nutrient loads, and internal ratios of nutrients when out of balance drive the waterbody toward eutrophication.

Thermodynamics drive important "triggers" which launch Harmful Algal Bloom events of which temperature is a very important driving factor.

The major factor in preventing Harmful Algal Blooms in Ponds, Lakes, Streams and Rivers is to control Phosphorous inputs. This is the first wave of defense in protecting your waterbody from highly invasive chemical inputs. Additional sources of Phosphorous is from wildlife, and birds which deposit organic material into the waterway.

Internal Nutrient Loading

In addition to External Nutrients the Phosphorous load in a pond, lake, stream, river, and reservoir is also influenced by "internal" sources.

Phosphorous can come from external sources as discussed above and also from internal sources which include fish waste, and Phosphorous which accumulates in bottom sediments. Agitation from bottom feeding fish, or mechanical disturbances from moving water can introduce nutrients which were "dormant" in the sediments. The release of these nutrients also stimulates algae, or Cyanobacteria growth rates and should be considered in a water management strategy.

Fish populations have dramatic impacts on Algae/Zooplankton/Fish relationships under the "energy cascade" model.

The variety and specific species of fish in a water body have significant impacts on water quality chemistry and nutrient loading. For example, if you choose a fish species native, invasive, or stocked which eats zooplankton, then algae and Cyanobacteria's natural predator is reduced in number giving Cyanobacteria a further edge toward increased growth events.

Fish species which are bottom feeders tend to disturb bottom sediments which releases nutrients (Phosphorous and Nitrogen compounds) from the hypolimnion layer into other depths and water levels within the water body also stimulating algae, or Cyanobacteria growth.

Restricting Nutrients to your Waterway:

The First line of defense is to prevent Phosphorous from entering your water body causing a "nutrient loading" event.

Chapter Five: Oxygen is Key

Healthy water bodies support a complex "food web" which spans the entire range of life forms including primary producers, primary consumers, primary and secondary predators ending with carnivores at the top of the food chain.

In ponds, lakes, streams, and rivers primary producers are aquatic plants (Algae, Diatoms, Cyanobacteria, and Macrophytes) which use photosynthesis to produce biomass and oxygen which is dissolved into the water column.

Primary consumers are usually zooplankton which feed on the algae (the primary producer). Biomass from the primary producer is a source of nutrients and feed larger animals up the food chain. Higher life forms like little fish feed on Zooplankton, in turn feeding small fish, which feed larger fish.

The "food web" is a trophic system defined by these stages from primary production to predation. Each "trophic" step from Primary Producer through Primary Consumer and on to predators each consume in their own metabolism about 20% of the available energy originally present in the primary producer biomass. Trophic steps are usually limited to 4 or 5 levels for this reason.

For example, solar energy driving photosynthesis in a water body produces a cascade of energy from Algae (a primary producer), to Zooplankton (primary consumer), to Small Fish (predator), to Large Fish (predator), or Humans (Carnivore). To produce 1 Kg of Salmon requires 5 Kg of smaller fish, which requires 25 Kg of still smaller fish, which in turn require 125 Kg of Zooplankton which requires 600 Kg of Algae (the prime producer). It takes a lot of algae to make some large fish.

Therefore, for purposes of illustration each Kg of Salmon consumed required 600 Kg of Algae to be produced, (in reality its much more). To support life on earth the aquatic plant kingdom must support an enormous amount of growth.

On average, Algae are coveted because of their rapid growth rates when conditions are right. Algae spend very little energy building "infrastructure" as terrestrial plants require. Algae have no roots, trunks, stems, or leafs putting all of their "energy" into creating biomass of incredible value. Beyond the needs of metabolism, and reproduction, acquired biomass from photosynthesis is in the form of valuable Proteins, Lipids, Carbohydrates, Anti-oxidants, Organic dyes, Enzymes, Vitamins, Phycobillioproteins and other metabolites of great value.

All higher life on Earth depends on "Primary Producers" through the Plant Kingdom which provide the essential nutrients and oxygen through photosynthesis for complex life to exist. There are essentially two forms of life on Earth: autotrophs, and heterotrophs. Autotrophs can produce their own nutrition directly from the environment (solar energy, mineral salts, CO2, and water), whereas heterotrophs (like us) must eat other organisms to attain complete nutrition. Because all higher life requires oxygen, all life on Earth depends on the Plant Kingdom for base production. Without plants, both terrestrial and aquatic, there would be no worms, insects, crustaceans, fish, birds, animals, or humans. Without the Plant Kingdom, the world doesn't eat or breathe.

Autotrophs, (the Plant Kingdom) on Earth produce the essential nutrition for the base of the food chain, and support and nourish all higher life forms through photosynthesis.

Oxygenic photosynthesis active in all terrestrial and aquatic species of algae (and Cyanobacteria) "oxidize" water and "reduce" Carbon Dioxide (CO_2), and through the Calvin-Benson Cycle, produce glucose as a basic building block for all proteins, lipids, carbohydrates, enzymes, vitamins and anti-oxidants. As if life needed to evolve a "battery" these simple sugars are used by cells as the basic building blocks of all higher bio-metabolites (cell metabolism net products) storing energy for cell life, reproduction and bioaccumulation. All Proteins, Lipids, and Carbohydrates are "built" by photosynthesis in the "light independent" processes described by the Calvin-Benson Cycle which shows how Carbon Dioxide (CO_2) is chemically reduced and "fixed" into molecules of glucose.

Cyanobacteria are masters of Oxygenic Photosynthesis. Harmful Algal Blooms (HAB) present a problem for eutrophic water bodies because of the enormous adaptability of the Cyanobacteria taxon to take over an aquatic ecosystem once conditions allow. Cyanobacteria can take hold.and can produce toxins which are dangerous to fish, amphibians, water fowl, farm animals, pets and people. The next chapter examines how Harmful Algal Blooms can be prevented, and treated.

Chapter Six: Treating Harmful Algal Blooms (HABs)

Algae, Diatoms, Cyanobacteria (Blue-Green Algae) and Macrophytes are the base of the food chain in normal and thriving aquatic systems active in ponds, lakes, streams, and rivers, serving a vital function as a primary producer of base nutrition and dissolved Oxygen.

"As the base goes, so the rest goes," is the basis of the "bottom up" approach to water management.

The internal conditions within your water body will dictate which algae or Cyanobacteria will be favored and has the greatest opportunity to dominate the waterway ecosystem.

There are several approaches to dealing with Harmful Algal Blooms (HAB) once they've occurred. These measures include different strategies to inhibit one or more of the necessary conditions needed by the Cyanobacteria or Algae to bloom.

There are several approaches to preventing the conditions and triggers which set a Harmful Algal Bloom in motion.

Nutrient and energy inputs are key to primary producers including Algae and Cyanobacteria in a aquatic ecosystem and are the first considerations.

Solar energy changes in intensity and duration as the seasons progress are conditions sensed by Algae and Cyanobacteria and induce very specific responses in those organisms to deal with, or take advantage of the changing conditions.

Daily variations from weather and micro-climates all affect algal growth rates from an Energy-In standpoint. One approach to treat a Harmful Algal Bloom is to block off or inhibit as much of the available light as possible from reaching into the water column.

Chemical approaches have been used as additives to color the water to be more opaque absorbing the solar energy coming into the water, slowing phytoplankton growth. As with all complex systems, alter one factor, and you'll alter others, so unintended consequences are observed when using additives. Despite the costs of these approaches, chemical additives are often used such as Copper Sulfate. Great concerns of increasing toxicity in the water and ecosystem from prolonged use of Chemical additives presents an exposure which is not recommended in this book.

Treating Harmful Algal Blooms

When conditions are favorable in bodies of water such as calm, warm, and nutrient rich, Cyanobacteria can take hold and grow very rapidly, inducing an exponential growth phase.

There are several approaches to dealing with Algal, and Harmful Algal Blooms (HAB).

The Most Important method of dealing with Harmful Algal Blooms is to prevent them from occurring in the first place. The most effective preventative measure is to "restrict" the availability of Phosphorous (nutrients) in your waterway.

Phosphorous and Nitrogen are the most important macro-nutrients and dictate the conditions for Algae, and Cyanobacteria.

Phosphorous control is the key. Because Cyanobacteria have so many "tools" in their capability just limiting Nitrogen inputs as a strategy is less effective than limiting Phosphorous which is the key driver. Cyanobacteria can often "fix" nitrogen (a vital macro-nutrient) from the atmosphere, instead of the water, which gives them an incredible advantage for surviving hostile changes in their water environment. When conditions in the waterbody are Nitrogen limited, such as when forms of inorganic Nitrogen are unavailable in the water, Cyanobacteria can manufacture Nitrogen compounds to continue to process Phosphorous. Cyanobacteria use this advantage in outcompeting other aquatic species when vital Nitrogen compounds are unavailable in a waterbody.

Cyanobacteria's ability to tap the atmosphere for a Nitrogen source converting Nitrogen into inorganic Nitrites and Nitrates which can be used for nutrition. Simply limiting Nitrogen in the water has little effect. The key is controlling (limiting) Phosphorous.

Nutrients in a water body can come from natural and man-made sources, and manifest from "internal" and "external" sources to your waterbody.

External sources which originate outside of your pond, lake, river, or reservoir include point, and non-point sources.

Point sources are easily identified as coming from a specific location and include Confined Animal Feeding Operations (CAFO), Wastewater Treatment Plants, Storm water run-off, Agricultural and Industrial activities which find routes for effluent from these activities to discharge into your waterway.

Non-point Sources of pollution include "land management" activities around the waterbody such as residential use of fertilizers, herbicides and pesticides. Other non point sources include water fowl, animal and pet waste, septic systems and general watershed runoff.

Controlling these sources of Phosphorous are the first line of defense in isolating your waterbody from a "nutrient" loading event. Methods of limiting external loading includes berms, and other barriers to channel water flow away from the waterbody, but this is often expensive and requires maintenance. Point source controls are regulated by DEQ and ODA under the Agricultural Water Quality Management Act (SB1010).

Practical methods of dealing with external sources include planting native aquatic plants (Macrophyte) around the perimeter to shore up soils.

Internal Nutrient Sources and Management

Internal nutrient loading can come from chemical processes in your waterbody reflecting higher pH from increased photosynthetic activity, as well as biological causes including burrowing animals, fish and amphibian movements as well as physical causes such as wind, sediment bubbling and use by water fowl and animals, (Cooke et al, 2005).

Internal nutrient "loads" are generated within a waterbody by nutrients being released from bottom sediments and excrement from fish, water fowl, and wildlife. The particular fish species which inhabit your waterbody have wide effects depending on the feeding habits of the fish, and the chemical composition of their wastes. The fish species Tui Chub, for example, have a particularly heavy Phosphorous loaded excrement and can load small lakes with Total Phosphorous (TP) levels which can ultimately trigger algal bloom events.

Lake and water managers often seek to identify the major fish populations and measure the effect of fish species on their internal nutrient load sources. This "top down" of "trophic cascade" approach seeks to identify fish species whether native, invasive, or populated and measure the effects of internal Phosphoric loading. Ponds, and lakes with fish species which feed on Zooplankton, the natural predator of Algae and Cyanobacteria, can also lower a waterbodies ability to consume Algae and

Cyanobacteria which is the natural way Algal populations are kept in balance.

In-Lake Management using Nutrient Limiting Techniques:

Once Phosphorous levels become elevated in your waterbody, one technique to "remove" Phosphorous from the water column is to add chemical flocculants. Flocculants, or Chelates cause a chemical precipitation by typically adding "alum," salts, calcium, iron, or other leaching agents which react and combine with Phosphorous in the water. These flocculants grab the Phosphorous out of the water by binding and begin to sink settling to the bottom of the water body removing Phosphorous from the water column. Although this approach may partially clear the water column for short periods of time it still builds up Phosphoric compounds in your bottom sediments which can accumulate and be discharged at a later time into your water increasing your internal loads.

When bottom sediments are determined to be heavy with Phosphorous and a source of internal Phosphorous loading another approach is to remove, or circulate the bottom layer of water which is most effected. This bottom layer or "hypolimnion" layer of water can be pumped out of the waterbody itself, or pumped to the surface to mix with other layers of the pond or lake in an effort to bring more Oxygen into the bottom layer of water.

If your bottom layer goes hypoxic then anaerobic bacteria can thrive which endanger your healthy aerobic bacteria.

Another approach to Phosphorous laden bottom sediments considered by lake managers is to actually dredge the pond or lake in an attempt to remove highly polluted bottom soils. This is a highly invasive approach and is seldom used as sediments need to be transported and can cause disruption to the aquatic ecosystem, but with very shallow ponds with very high levels of Phosphorous it has been considered in extreme cases.

Another approach to lowering the internal Phosphorous loading is to dilute the water body with the introduction of new water. This approach is also expensive, and is somewhat invasive as aquatic plants, microbial organisms, and fish are very sensitive to sudden water quality changes. It's not just the change that occurs in water quality which can stress aquatic life, it's often the rate-of-change which can be equally shocking.

Water Mixing

Stagnant unmoving waters are another "trigger" mechanism which can promote algal blooms. When waters are stagnant they can begin to stratify with layers of different temperatures and Oxygenation. Techniques to move, and mix the water layers include fountains, circulation pumps and aeration. The smaller the body of water the more effective these techniques. Fountains can be an effective aeration device especially effective for top layers of water depending on power, the rate of water exchanged, and the depth at which the feed water is sourced.

Fountains which only pump from the surface waters are not very effective at distributing the Oxygenated water at significant depths in your waterbody. Fountains which pump water from deeper layers of your waterbody serve to increase the overall distribution of dissolved Oxygen in part through the mechanical mixing which occurs when water is pumped from the bottom. Water pumped is replaced and a basic circulation can be created.

Circulation pumps are another technique to mix water layers and can be configured to pump surface water to the bottom, or bottom water to the surface. Most effective mixing from a dissolved Oxygen standpoint is to bring the water from the surface. Circulation pumps not only help distribute Oxygen throughout the water column but also increase turbidity and suspended particles in the water. Although this increases the Phosphorous levels at different depths it also "exposes" Phosphorous and Phosphoric compounds to dissolved Oxygen which chemically react binding up Phosphorous and rendering them "unavailable" for use by Cyanobacteria and Algae.

Increasing turbidity through mixing also blocks light from penetrating very deep into the water further limiting photosynthetic bacteria from growing, balanced against increasing the short term availability of nutrients.

Aeration

Aeration is the act of pumping air into the water column. Injecting air directly into the water, especially at depth is highly effective at delivering Oxygen to all layers of your waterbody. Aeration is so important as a method of preventing and treating Harmful Algal Blooms (HAB) that the next chapter is devoted to the subject.

Water temperature is another major "trigger" for algal blooms. Thermodynamically, warmer water generally assists metabolism, respiration, reproduction and photosynthesis in aquatic plants with some important limits. Each taxa of Algae, and Cyanobacteria, have their own "preferred" temperature at which the taxa thrives and can grow exponentially. Cyanobacteria are often more temperature tolerant than other algae and aquatic species offering another evolutionary advantage.

Temperature is vitally important for a particular species with sensitivities often measured in the tenths of degrees. Ponds, lakes, streams and reservoirs are particularly susceptible to temperature pollution with their often shallow depths causing over heating in the water column and stratification in deeper lakes.

Solar energy at peak beams 1,000 watts per square meter on water bodies which can provide a huge solar gain. One acre of water surface (over 4,000 square meters) exposed to the sun at peak is an input of 4 Megawatts of total energy. Over one hour of time that's the equivalent of 4,000 Kilowatt-hours (kWh) of energy. Although some energy is reflected from the surface, the solar gain is critical for photosynthetic organisms.

Except for very shallow water bodies, water temperatures are cooler at depth. One approach to lowering temperature in a pond or lake is to pump water from the bottom cooler layers to the surface to mix the cooler water with the warm surface water to lower the over all temperature and impede algal blooms. This approach can be potentially dangerous for cold water fish which might inhabit your lake such as trout who need cool temperatures of the deeper layers to be healthy. Warm water species are much more tolerant to this approach.

Climate change affects ponds, lakes, streams, rivers and reservoirs with ever increasing average temperatures. Increases in water temperature put pressure on "Green" algae, and Diatoms which are generally healthy for your waterways - taking up Phosphorous and Oxygenating the water. Higher temperatures stress Green algae and Diatoms and opens the door for highly adaptive Cyanobacteria taxa to find a hold. Climate change has increased the frequency of storm events which can wash significant amounts of Phosphorous and Nitrogen compounds into your waterbody greatly increasing the path towards eutrophic conditions likely to develop.

Factors including climate change, increased discharge of Point sources of Phosphorous near your water body, and the actual species which inhabit your water ecosystem present large challenges for Lake and Water Mangers.

Water mangers develop management plans to keep Total Maximum Daily Load (TMDL) inline with Federal, State and local standards. Therefore, water managers seek to implement "internal" and "external" management plans to mitigate and keep TMDL levels under limits.

Chapter Seven: Aeration and Circulation Technology

The purpose of aeration in water management is to increase the content of dissolved Oxygen in the water and to mix different layers of water in one step.

There are various systems which can provide some level of aeration including injecting air into the water column, mechanically mixing or agitating the water and fountains which spray water into the air atomizing and Oxygenating the water.

Aeration improves fish and other aquatic animal habitats, prevents fish kills, and improves the water quality by increasing the levels of dissolved Oxygen in the water, vital to healthy aquatic ecosystems.

In natural waterways wind action and natural sources of agitation make atmospheric "diffusion" of Oxygen the principle mechanism of external aeration. Internal sources of dissolved Oxygen are produced from primary producers (Algae, Diatoms, Cyanobacteria, and Macrophytes). Together, phytoplankton and the atmosphere provide a healthy aquatic system with dissolved Oxygen from 5 mg/Liter to 10 mg/Liter in most "normal" waterways.

The dominant method of aeration for treating deeper ponds, lakes, rivers, and reservoirs which are suffering from hypoxia use air-injection diffusion systems. An air-injection system uses a compressor onshore, or floating on a suitable platform, to pump air through a tube into a receiver set at depth. The receiver at the endpoint is either perforated pipes, or micro-diffusers which are a high density of micro holes intended to produce micro-bubbles when the compressors are running. When bubbles rise from the Diffuser their enormous collective "surface area" displaces an enormous volume of water and causes a vertical circulation of water through the various water depths.

Thousands of micro-bubbles rising from depth act to increase dissolved Oxygen levels at the deep hypolimnion layer and to drive this gentle mixing of water layers, bringing bottom to top, and top to bottom.

Most Oxygen gas diffusion into a waterbody happens at the surface with the interaction of the water surface with the atmosphere. Aeration brings deep water up where it travels laterally across the surface where most of the Oxygenation occurs. After some time the surface water begins to sink replacing the water volume being lifted and a "turn over" of the lake water is induced.

This "turn over" effect is a very efficient way to Oxygenate the deep water layers of your water body.

Positioned deep in the lake, for example, the rising air bubbles cause water in the hypolimnion layer (the coldest bottom layer of water) to also rise, pulling the cold water into the warmer epilimnion layer (surface layers).

Rising bubbles pull cooler water from depth to the surface where it travels laterally, then sinks, bringing warmer waters down to the cooler lower layers inducing a vertical circulation pattern helping to keep the thermal layers mixing. This helps Oxygenate the bottom layers both from the aeration bubbles (about 5% Oxygen transfer) and from water being brought down from the surface with the highest dissolved Oxygen content (about 95% Oxygen transfer).

Invasive Cyanobacteria prefer still waters. Aeration causing mixing and Oxygenation helps discourage the Cyanobacteria from taking hold and forming colonies.

Many dynamics are now involved as nutrients from the bottom layer which can spike algal growth are mitigated by the presence of Oxygen which binds with the Phosphorous (Chemical Oxygen Demand) making it unavailable for algal growth. This push-me pull-you dynamic continues until the waterbody reaches a dynamic equilibrium.

Another "Top down" approach is to use Mechanical Axial-Flow pumps which are designed to "push" surface water down to lower levels in your waterbody. The Axial Flow pump uses a floating platform with a large diameter propeller mounted below the platform facing down. A power supply with an electric motor and gearing drive the underwater propeller or paddles which, like your air fan, produce a flow of fluid as in this case, water downward. This brings Oxygenated water from the surface aerated by exposure to the atmosphere down to the hypolimnion layer (bottom zone) where lake water is Oxygen poor by comparison.

As water is "pushed" down by the blades, Oxygen-poor water from the bottom is brought to the surface where diffusion from the atmosphere Oxygenates the water. In this way dissolved Oxygen and temperature is more evenly distributed throughout the waterbody.

This is a very effective way to combat temperature stratification in your waterbody which if left to thermally stratify can induce bottom layer water hypoxia and anoxia.

Fountains, impeller-aspirators and pump-and-cascade systems are also used to interact water with air for Oxygenation. These are effective approaches for Oxygenating surface and shallow depth water, but less effective for reaching deeper water where hypoxic conditions can cause an Anaerobic condition to flourish which pulls Oxygen out of the water. For good aquatic health it's important to implement plans which increase the dissolved Oxygen content found at the bottom of pond, lake, river, and reservoir water bodies.

Aeration and circulation can also be highly effective in preventing wintertime "fish kills." Ice covered lakes and ponds keep air from interacting with the surface water where Oxygen can diffuse into the water.

Aeration by pumping air down into the waterbody with a diffuser to produce micro-bubbles is an effective way to keep water ecosystems well Oxygenated during winter months.

Cyanobacteria prefer calm and static waters. Aeration moves the water as it "turns over" during which operation this keeps potentially Toxic Algae strains abated.

Chapter Eight: Managing Ponds, Lakes, Rivers and Reservoirs

The objective of managing ponds, lakes, streams, rivers, and reservoirs is to protect and nourish a healthy water body which supports a wide diversity of life including algae, diatoms, Cyanobacteria, macrophytes, zooplankton, amphibians, crustaceans, fish, insects, birds, animals, and people.

Different bodies of water including ponds, lakes, streams, rivers, and reservoirs have different dynamics and local conditions. Spanning physical, chemical and biological factors experiencing Internal and External sources of loading and stress, management plans must consider a complex system in developing a comprehensive management plan. Understanding the specific conditions of the waterbody, and how these conditions change, is the foundation of establishing a comprehensive protocol.

Ponds

Ponds are typically shallow (less than 15 feet) and fairly uniform in the distribution of temperature throughout each depth. This lack of "thermal stratification" means the bottom layer of the pond (the Hypolimnion layer) closest to the bottom is close to the temperature of the surface.

The action of the winds on the water surface typically keep a small pond well mixed and sampling of dissolved Oxygen will be roughly consistent at all depths. Ponds, or small lakes with large areas of warm and shallow water usually have large sediment collections at the bottom which allow relatively easy mixing into the water column from the mixing action of the wind. This stimulates algae, and Cyanobacteria growth by providing constant nutrient introduction with this "internal" loading.

Because shallow ponds, and small lakes have water mixing happening through wind action and water movement throughout the water limiting "external" sources of Phosphorous are less effective at slowing algal growth as "internal" loading from sediment mixing stimulates algal growth.

Lakes:

Lakes range from fractions of acres to many thousands of acres and typically have areas with depths deeper than 15 feet. Shallow lakes are very nutrient rich typically from sediments mixing into the water column with water movement from wind action, bioturbation and physical bubbles emerging from Methane and Hydrogen Sulfide gasses escaping from decomposition.

Shallow lakes are often defined as water bodies where the phototrophic zone extends down to the bottom and with little thermal stratification.

Deeper Lakes have lower water temperatures at depth and have a thermocline layer (the boundary layer) where there is a large difference in water temperatures between layers. The bottom layer or hypolimnion layer is usually thermally stratified and where the lowest temperatures in the lake are found. Henry's Law of gas diffusion suggests that because of cold temperatures Oxygen could be readily soluble and we can expect the colder water can support higher levels of dissolved Oxygen. However, if lakes become eutrophic the hypolimnion layer overloads with nutrients and biomaterial becoming anoxic and supports anaerobic organisms where dissolved Oxygen can reach nearly zero.

Water quality managers examine the external sources of Phosphorous and internal sources of nutrient loading. Internal sources include sediments, vegetation and fish species which influence algal growth in water bodies. Zooplankton are the primary consumer which eats algae and keeps a healthy waterway in balance.

Bio-manipulation techniques are considered by water managers to limit invasive species of fish which graze on Zooplankton the natural predator of algae and Cyanobacteria. Bottom feeders like carp, shad, and bullheads feed on Zooplankton and this promotes phytoplankton like algae and Cyanobacteria to grow.

Typically, ponds and lakes have either one of two likely conditions. Either a pond or lake is dominated by Macrophytes (large aquatic plants) and are mostly clear in water clarity, or you have a cloudy, colored eutrophic water which has an algal bloom, is turbid, and has an imbalance or overloading of Phosphorous. As macrophytes are in competition with Algae and Cyanobacteria it is unusual to find a shallow lake free of both aquatic plants and algae simultaneously.

Streams:

Streams are characterized as creeks or small flows of water usually feeding into larger and larger flows which become rivers. Streams can also empty into lakes, reservoirs and water sheds in general.

Streams support complex ecosystems which are rooted in aquatic plants both macro and micro (Algae, Diatoms, and Cyanobacteria) as primary producers.

Water quality in streams can vary greatly and are largely dictated by water flow rates. Fast moving water tends to Oxygenate the water, keeping aerobic plants, amphibian and fish populations well Oxygenated and able to thrive.

Slower moving, or unmoving parts of a stream can have "micro" climate effects in the local water column of more hypoxic conditions where Cyanobacteria are most likely to take hold.

Rivers:

Rivers usually have fast water flows with areas of lower and even slow flow, as well as side pools and channels where algae or Cyanobacteria can bloom.

Organic inputs from water fowl, animals, and human sources like septic systems, and agricultural and storm event runoff can fuel localized Algal Blooms which can later spread to other parts of the river down stream.

Water quality in a river depends on where the sample is taken as a wide range of conditions can exist depending largely on the speed and turbulence of the water flow.

Reservoirs:

Reservoirs face many challenges due to their physical shape usually elongated, and require a limited ecosystem due to organic sources of dangerous pathogens such as e-coli bacteria, listeria, and cholera, and are vulnerable to frequent runoff events from increased storm activity, longer periods of drought, and lower snowpacks in the water shed.

Reservoirs are often located "down stream" in a watershed and may receive much more effluent than lakes for example. Their often elongated shape gives a reservoir a much larger exposure to shallow waters which interact with sediment layers increasing mixing and internal nutrient loading.

Water Management Plans

Harmful Algal Blooms (HAB) represent an ever increasing threat to all healthy ponds, lakes, streams, rivers, and reservoirs.

Water managers seek to develop water management plans which define the conditions within waterbodies through out the year, and an action plan to deal with toxicity arising from a Harmful Algal Bloom (HAB) event.

External factors such as climate change, and increased intensity of industrial and agricultural runoffs increase the pressure on waterbodies from a chemical standpoint.

The cornerstone of pond, lake, river and watershed management is to prepare a plan of action and to implement that plan based on the changing conditions of the waterbody.

All water bodies from mud puddles, ponds, lakes, rivers, reservoirs, estuaries, wet lands and oceans have complex ecosystems. It is important to not let one action of implementing a management plan targeting one aspect of water quality have unwanted and unintended impacts on another.

Water managers must make complicated calculations, and considerations regarding different measures of dealing with Harmful Algal Blooms (HAB) and how the waterbody is affected by preventive and treatment measures taken.

There are three general regions to water management regarding aquatic ecosystems. The first is the surrounding wetlands and littoral zones and the sediments which occur from these surface zones.

The next biotic system is the pelagic zone and includes most of the water column where primary producers (algae, Cyanobacteria and Diatoms) supply primary consumers such as Zooplankton with base nutrition.

This is the zone where photosynthetic organisms and aerobic bacteria in a healthy water body inhabit.

The third biotic region is the benthic bottom layer zone. The bottom layer hypolimnion zone is one of the most fragile of the entire ecosystem and is vulnerable to chemical (and organic) inputs into your waterbody.

Each of these three zones interact and actions to modify one aspect of one zone can, and will, effect all other zones. It is imperative for watershed managers to take a holistic approach to treating Harmful Algal Blooms (HAB) and seek solutions which address each zone and how they interact.

For example, if a lake is over grown with Macrophytes (large aquatic plants) and a program is implemented to remove these plants, then nutrients formerly taken up by these plants are then more available for microbial growth and algal blooms can be promoted. Removing Macrophytes also weaken the surrounding soils which can promote leaching of Phosphorous further adding to the "internal" loading occurring in your lake, pond, or waterbody.

Adding colorants in an attempt to limit solar energy penetrating the lake can also suppress beneficial bacteria and alga. Adding colorants may also have cumulative toxic effects which will manifest at a later time. The use of herbicides, and algaecides may provide short term benefits, but will have effects on other life in the aquatic ecosystem and therefore doesn't present a long term sustainable solution to Harmful Algal Blooms (HAB) mitigation.

Lake Treatment Example:

Consider an example lake with a surface area of 13 acres and a maximum depth of 34 feet for our test case. Consider this lake built as a retention pond and surrounded by fertilized and "managed" landscaping and residential use. Part of the Lake borders a golf course which is also a significant source of nutrient loading.

At 13 acres our sample lake is located in a Southern state and is seeded with fish for recreational fishing enjoyed by the locals.

With ample sources of external loading from the fertilizers and the golf course, and internal loading from the fish excrement (Shad) the warm weather of the Summer causes an algal bloom which "breaks bad" by becoming a Harmful Algal Bloom (HAB)

Populated with fish our lake begins to experience "fish kill" events in late Summer/Fall as a run away effect from algal blooms and massive algal deaths creates an abundance of dead biomaterial in the water for aerobic bacteria to feed. This bacteria pulls Oxygen out of the water through Biological Oxygen Demand and the hypoxia results in fish kills.

The Summer fueled growth of Algae, Cyanobacteria, and Macrophytes begins to die off when more inclement weather and seasonal changes leave large amounts of organic matter suspended in the water.

As organic material sinks through the water column it is consumed by aerobic bacteria which greatly increase the Biological Oxygen Demand (BOD) pulling Oxygen out of the water. Other than mild mixing through wind action on the lake surface, the bottom of the lake begins to become anoxic with an abundance of organic matter being decomposed anaerobically resulting in noxious gas (Hydrogen Sulfide) being released.

The lake begins to stratify thermally, and levels of dissolved Oxygen at the bottom layer of water (hypolimnion) zone become anoxic and won't support fish life. Anoxic bottom water layers promote the growth of "anaerobic" bacteria beginning to thrive emitting hydrogen sulfide creating a "dead" zone at the bottom of the lake. If left unabated, this dead zone will begin to migrate upward ending with a dead body of water except for dominant Cyanobacteria which may be toxic destroying the normal ecosystem, and producing a source of toxicity for other wildlife, farm animals, pets, and people.

To bring this sample lake back into balance and able to support a healthy and diverse aquatic ecosystem the technique of Aeration is used.

Compressors can be utilized at about 2.5 Horsepower (Hp) per 10 acres of Lake. Situated usually onshore, these compressors pump 15 Cubic Feet per Minute (15 CFM) through tubing to diffuser bars are submerged and placed at the deepest part, or near deepest part of the Lake Aeration diffusers are perforated to produce micro-bubbles which bubble up from the diffuser pad. As the bubbles ascend water is displaced upward causing a "draw" of water to replace water being lifted. This forms a circulation pattern which brings low Oxygen bottom water to the surface.

The atmosphere-water boundary is where the most Oxygen is dissolved into the water (other than internal photosynthesis) and as the water reaches the surface from the bubbling it tends to travel laterally across the surface of the lake absorbing Oxygen.

As that water eventually sinks it bring Oxygenated water back down to the bottom layers (hypolimnion) and begins treating the Anaerobic bacterial, and an environment begins to become Aerobic and able to support bottom dwelling benthic life.

Aeration diffusers (large bubble stones) are placed at or near the bottom of the deepest part of lake, or placed along the center line of the lake evenly separated. As air is pumped from the surface down to the diffusers the bubbles rise and induce the circulation of bottom water toward the surface.

It is important to consider the start up phase of this physical gear. As soon as the aeration compressors are engaged there will be a period of increased turbidity which may be stressful to fish. However, when the lake begins to adjust to the increase in Oxygenation then food webs begin to recover to reach a dynamic equilibrium which is healthy for your aquatic ecosystem.

Methods of Treatment

Understanding a waterbody's water quality is vital for formulating the management plan designed to protect aquatic ecosystems, and discourage conditions which promote Harmful Algal Blooms (HAB). Study, define, understand and implement your plan for maximum effect.

Measure water quality factors including Chlorophyll-a, pH, Total Phosphorous (TP), Total Nitrogen (TN), Chemical Oxygen Demand (COD), Biochemical Oxygen Demand (BOD), turbidity, and color to reach your Total Solids Index (TSI), to determine the water bodies Total Maximum Daily Load (TMDL). Measurement regimens and protocols should be constructed to sample data over time.

Document the External, and Internal sources of Phosphorous nutrient loading.

For larger and deeper lakes (deeper than 15 feet in part) identify Phosphorous sources from Point-source, and Non-Point sources to consider how to treat the original effluent, or divert runoff from entering the waterbody.

Devise the method best suited to the particular situation to limit nutrients, increase mixing, and Oxygenation. Choices include:

Phosphorous Inactivation through chemical precipitation such as adding Alum or Chelate. (Undesirable for cost, and potential toxic impacts).

Dredging Sediments. (Undesirable for cost, disposal issues, and toxic impacts due to turbidity and mixing)

Hypolimnion Water removal. (Undesirable for cost, disposal, and impacts on species)

Dilution and flushing. (Undesirable for cost, and impacts on species).

Bio-manipulation. (Invasive, and can have unintended consequences on other trophic levels of the aquatic ecosystem).

Hypolimnion Layer Aeration - is the Best Approach of all.

The most effective treatment to protect your water body from becoming hypoxic, chemically imbalanced, and potentially capable of triggering a Harmful Algal Bloom (HAB) is Aeration.

Aeration is not a silver bullet, but is the single most effective method which can achieve the greatest long term effects with no toxicity, or invasive manipulation. Oxygen levels at various depths are the most illustrative metric and allows water managers to employ methods which increase water layer mixing (to minimize stratification) and increase the dissolved Oxygen at the bottom of the aquatic ecosystem.

The Hypolimnion (bottom) layer, the water column, and the surface water health are each, and all, vital for balanced healthy aquatic ecosystems. Each of these vital zones are improved with the thoughtful use of Aeration technology increasing water layer mixing and Oxygenation.

Oxygen is the key to life on Earth. The introduction of Oxygen into the entire food web in your pond, lake, river, or reservoir will promote desirable aquatic, fowl, and animal life and health, while simultaneously discouraging undesirable species such as toxin producing species of Cyanobacteria.

Toxic algae in the form of Harmful Algal Blooms (HAB) can be prevented and treated with proper management and methods.

Proper use of aeration technology is the most effective way to treat, and prevent Harmful Algal Blooms (HAB) in ponds, lakes, streams, rivers, and reservoirs.

About the Author

Christopher Kinkaid

Christopher (Toby) Kinkaid, originally from Portland, Oregon, is the founder of **Solardyne.com**, **SolarQuote.com**, and **AlgaeToday.com**, and has worked in clean energy technology for over three decades.

Christopher Kinkaid, is the inventor of the "Helyx" Vertical Axis Wind Generator, the "Mariposa" Non-imaging solar concentrator PV module (continuous operation at Sandia National Laboratory since 1994), the Solar Demultiplexer optical solar concentrating lens (Dr. James/Sandia National Laboratory 1991), and the inventor of the original "Solar Power Pack" (Mother Earth News, "Littlest Utility" June/July, 2001).

Christopher Kinkaid, has lectured on clean energy technology around the world including "APEC", Bangkok, Thailand, 2003, "Energy Solutions World", Tokyo, Japan, 2003, The International Biomass Conference (IBC), 2010, Minneapolis, MN, and the

Algal Biomass Organization (ABO) Conference, 2010, Phoenix, AZ.

Christopher (Toby) Kinkaid, has appeared in interviews on KOIN TV, KGW TV, and "Sustainable Today" produced in Oregon, and has served on the board of directors for the National Hydrogen Association, in Washington D.C., 1993, the Japan Satellite Communications Company (JCNET), Fukuoka, Japan, 1994-95, and Algaedyne Corporation, St. Paul, MN, 2010-2013.

Christopher Kinkaid, presently serves as CEO of Solardyne, LLC in Portland, Oregon, where he continues his work in Solar, Wind, and Biomass Technology applications, research, and development.

Made in the USA
Middletown, DE
06 July 2020